MILITARY JOBS

DEMOLITIONS EXPERTS

▸ **What It Takes to Join the Elite**

TIM RIPLEY

Cavendish Square
New York

Published in 2016 by Cavendish Square Publishing, LLC
243 5th Avenue, Suite 136, New York, NY 10016

© 2016 Brown Bear Books Ltd

First Edition

Website: cavendishsq.com

This publication represents the opinions and views of the author based on his or her personal experiences, knowledge, and research. The information in this book serves as a general guide only. the author and publisher have used their best efforts in preparing this book and disclaim liability rising directly or indirectly from the use and application of this book.

CPSIA Compliance Information: Batch #WS15CSQ

Library of Congress Cataloging-in-Publication Data

Ripley, Tim.
 Demolitions experts : what it takes to join the elite / Tim Ripley.
    pages cm. — (Military jobs)
 Includes bibliographical references and index.
 ISBN 978-1-50260-518-4 (hardcover) ISBN 978-1-50260-519-1 (ebook)
1. Demolition, Military. 2. Explosives, Military. 3. Military engineers—United States. 4. United States. Army—Vocational guidance. 5. Military engineering—United States—Vocational guidance. I. Title.

UG370.R87 2015
358'.2302373—dc23

2014049222

For Brown Bear Books Ltd:
Editorial Director: Lindsey Lowe
Managing Editor: Tim Cooke
Children's Publisher: Anne O'Daly
Design Manager: Keith Davis
Designer: Lynne Lennon
Picture Manager: Sophie Mortimer

Picture Credits:
T=Top, C=Center, B=Bottom, L=Left, R=Right

Front Cover: U.S. Department of Defense
All images U.S. Department of Defense except:
Getty Images: Getty Images News 40; Robert Hunt Library: 6; U.S. National Archives 7.

Brown Bear Books has made every attempt to contact the copyright holder.
If you have any information please contact licensing@brownbearbooks.co.uk

We believe the extracts included in this book to be material in the public domain.
Anyone having any further information should contact licensing@brownbearbooks.co.uk.

All rights reserved. No part of this book may be reproduced, stored in a retrieval system, or transmitted in any form or by any means, electronic, mechanical, photocopying, recording, or otherwise, without the prior written permission of the copyright holder.

Manufactured in the United States of America

CONTENTS

Introduction 4

SELECTION AND TRAINING
History 6
What It Takes 8
Individual Selection 10
House-Entry Training 12
Battlefield Training 14
Sabotage Training 16
Underwater Training 18

SPECIAL SKILLS
Assault Teams 20
Clearing Obstacles 22
IED Destruction 24
Sabotage 26
Underwater Assault 28

SPECIAL EQUIPMENT
Explosives 30
House-Entry Equipment 32
Obstacle Clearing Equipment 34
IED Disposal Equipment 36
Armored Vehicles 38

IN ACTION
Zhawar Kili Caves, 2002 40
Baghdad, 2003 42
Helmand Province, 2009 44

GLOSSARY 46
FURTHER INFORMATION 47
INDEX 48

MILITARY JOBS

INTRODUCTION

Whether planting explosives under enemy fire, disarming potentially lethal improvised explosive devices (IEDs), or destroying bridges or buildings, demolitions experts need steady nerves and good concentration.

Demolitions are one part of the job of a combat engineer. Combat engineers also build roads and bridges, trenches, bunkers, and other defenses, but this book concentrates on their work with explosives.

In modern warfare, combat engineers use explosives for tasks from clearing a path through a minefield to blasting open the door of a house during a hostage rescue mission. They also use their expertise to prevent explosions by defusing mines and IEDs. The Army and the Marine Corps train their own engineers, but they all have one thing in common. As well as being expert engineers, they are also effective warriors, trained to be able to fight alongside other soldiers in any theater of war if the situation requires them to do so. Special Forces such as the US Navy SEALs also learn demolition skills.

▶▶ US Marine combat engineers take cover behind a shield, having blown a hole in a wall during training.

DEMOLITIONS EXPERTS

MILITARY JOBS

HISTORY

Demolitions experts appeared in the seventeenth and eighteenth centuries. Engineers dug trenches called "saps" to get close to castles or towns. Then they used gunpowder to blow holes in the walls.

US troops look on as a blast destroys caves used by the Japanese during the battle for the island of Okinawa in 1945.

In many armies, combat engineers are still known as "sappers." The trenches for which they are named protected them from enemy fire as they placed gunpowder against a wall. Sappers also sometimes dug under walls to make them collapse.

DEMOLITIONS EXPERTS

IN ACTION

On D-Day, June 6, 1944, US Navy and Army demolitions experts landed on the beaches of Normandy a few minutes after the first assault troops. Under heavy fire, they placed charges on steel obstacles and blasted paths onto the beaches. They had also played a key role in reconnaissance of the beaches in the weeks before the landing.

◀◀ US demolition engineers prepare fuses for an explosive charge in North Korea during the Korean War (1950–1953).

The trench warfare of World War I (1914–1918) saw growing demand for combat engineers. They dug tunnels to plant mines beneath enemy positions.

World War II

Modern-day US demolition units have their origins in World War II (1939–1945). In the Pacific, they played a key part in the amphibious campaign of the US Marine Corps. The Marines used demolitions experts to blow holes in beach defenses on Japanese-held islands. The engineers swam ashore to plant explosive charges. These underwater demolition teams were the predecessors of today's US Navy SEALs.

Meanwhile in Europe, US Army combat engineers were involved in the D-Day landings in Normandy in June 1944. They developed innovative tactics and technology to blast breaches in Hitler's coastal defenses.

MILITARY JOBS

▶▶ WHAT IT TAKES

Military demolition work is not for the fainthearted. It requires a combination of brains, strong nerves, and brute strength. Demolitions experts lead the way into battle, so they have to put themselves in harm's way.

Planting and setting off explosive charges requires a thorough knowledge of the science of explosives. Soldiers have to calculate how much explosive material to use to achieve the desired destructive effect. Too much explosive, and they and their colleagues might be in danger. Too little, and the target might not be destroyed. In action, the failure to destroy a key structure, such as a bridge, can be disastrous. Combat engineers cannot afford to make mistakes.

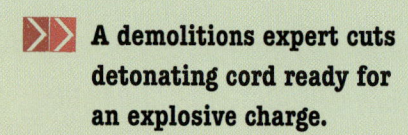

▶▶ A demolitions expert cuts detonating cord ready for an explosive charge.

DEMOLITIONS EXPERTS

Steady nerves are needed for demolition work close to the enemy. Carrying explosives into battle is inherently dangerous. To breach minefields, demolitions experts have to be in the first assault wave. They are on the receiving end of heavy enemy fire.

Physical Strength

Military demolitions experts need to be physically and mentally tough. They must work outdoors in all types of weather or on exposed beaches. They have to carry all their explosives and detonating gear with them into battle.

Marines shelter behind a shield as they destroy a mine. Demolitions experts often operate close to explosions, and under enemy fire.

EYEWITNESS

"The ideal engineer is a composite. He is not a scientist, nor a mathematician, nor a sociologist, nor a writer, but he may use the knowledge of any or all of these disciplines in solving engineering problems."

—Prof. N. W. Dougherty, University of Tennessee

MILITARY JOBS

▶▶ INDIVIDUAL SELECTION

Using explosives requires extensive training. Combat engineers learn a full range of skills. They have to be both technical and tactical experts, so they can use explosives in a wide range of battlefield situations.

Demolitions experts prepare a charge to destroy a simulated obstacle during an exercise at the Army Engineer School.

To qualify for training, combat engineers have to get a high score on the Armed Services Vocational Aptitude Battery (ASVAB) tests. Combat engineer training lasts fourteen weeks. Recruits learn to

DEMOLITIONS EXPERTS

Combat engineer recruits learn to prepare explosives during a class at the Army Engineer School.

build and destroy bridges and to create or breach obstacles, as well as general engineering principles. They learn to place and detonate explosives, to handle explosives safely, and to prepare and install firing systems for explosives.

Flexible Approach

During their training, combat engineers learn that they are "Warriors Always." Their motto is "Let Us Try." It is a reminder that they need to use flexibility and ingenuity to overcome the challenges they will face.

IN ACTION

The US Army Engineer School is based at Fort Leonard Wood, Missouri. It is open to both male and female recruits. However, although female recruits can be fully trained, they cannot serve as combat engineers. Women are not allowed to be assigned to frontline combat units. That situation may change as the US government reviews the role of female soldiers.

MILITARY JOBS

▶▶ HOUSE-ENTRY TRAINING

Gaining entry to terrorist buildings and vehicles is a key skill of the elite Special Forces. Units such as Delta Force and SEAL Team 6 undergo extensive training in blasting their way into hostile locations.

▼ An assault squad provides cover as an engineer places an explosive charge on the door of a "killing house."

Special Forces operatives practice on specially built structures known as "killing houses." They learn how to use small plastic explosive charges to blow

DEMOLITIONS EXPERTS

Combat engineers duck behind their shield as they blow open a door during a training exercise.

open windows or doors to allow a ground assault. They are also taught how to use more powerful charges to blast breaches in walls or roofs, so an assault team can abseil into a buiding from the roof.

Psychological Advantage

As well as practical skills, operatives are taught to move fast to exploit the shock caused by a sudden explosion. They try to enter the target building or vehicle as soon as the demolition charge is detonated so they catch the enemy while they are disoriented and demoralized. This gives the attackers an advantage, allowing them to capture targets or release hostages virtually unopposed.

EYEWITNESS

"Speed is the most significant factor in every type of breaching. Every time the squad is stalled because of a breach, it is placed in a vulnerable position. Breaching swiftly and effectively is necessary for the squad to maintain momentum."

—*Infantry Squad Tactics, Marine Corps Gazette*, 2005

MILITARY JOBS

▶▶ BATTLEFIELD TRAINING

In battle, demolitions experts blast safe routes through enemy minefields or other obstacles such as barbed wire. They must work quickly.

Assault troops may be exposed to enemy fire while they wait for demolitions experts to clear an obstacle. That puts extreme pressure on demolitions engineers trying to disable dangerous explosives.

▼ A combat engineer studies a landmine during training to assess its stability so it can be disarmed.

DEMOLITIONS EXPERTS

▼ **During a drill in clearing minefields, an Army combat engineer drags a grapnel hook on a rope over the ground to check for trip wires.**

In training, demolitions experts learn how to identify enemy minefields and obstacles quickly. They also learn how to position charges to have maximum effect, but in a way that does not endanger their own troops.

Combined Arms

Combat engineers train with infantry, armor, and artillery units to learn how to coordinate breaching operations. They approach the breaching point under the protection of other combat units. After a breach is made, the combat engineers ensure the safe passage of their assault force. They mark safe routes so vehicles and troops stay on the cleared tracks.

IN ACTION

For mine detection in dangerous or inaccessible areas, combat engineers call on the Engineer Detachment (Canine). The unit trains dogs to sniff out hidden explosives or buried mines. The dogs are trained to remain calm under gunfire. They can help locate mines for destruction or can guide a patrol on a safe path through a minefield.

MILITARY JOBS

▶▶ SABOTAGE TRAINING

Destroying equipment and structures is a specialized skill that requires intensive training. It can create obstacles in the path of an enemy advance, or can deny the enemy equipment or resources.

▼ Demolitions experts blow the wings off an abandoned C-130 Hercules aircraft in Iraq in 2007.

Combat engineers are taught how to destroy bridges and buildings. They learn how to place demolition charges at structural weak points, so the minimum amount of explosives will bring the structure down.

DEMOLITIONS EXPERTS

Demolition teams are taught how to work with infantry protection troops so a demolitions operation can be carried out close to the front line. Demolition teams also learn to destroy bridges at the last minute in case of a sudden retreat. They give friendly troops as much time as possible to cross before detonating the explosives and stopping the enemy advance.

Causing Damage

Combat engineers also learn to use explosives to destroy enemy equipment. They set out to cause maximum damage. Different factors influence the type of explosive to be used and where it is placed. Arms caches, vehicles, industrial plants, and other equipment are destroyed using techniques developed for the specific target.

EYEWITNESS

"You are an engineer. You are going to build bridges and blow them up. You are going to build roads, airfields, and buildings. You are going to make sure that our own troops move ahead against all opposition. You are an engineer."

—US Engineer Soldier's Handbook, 1943

▶▶ Two Marine engineers place plastic explosives on a tire rim during a demolition exercise.

MILITARY JOBS

▶▶ UNDERWATER TRAINING

Using explosives underwater is the job of naval Explosive Ordnance Disposal (EOD) technicians. It is also a skill of the Navy SEALs. They prepare for amphibious landings by destroying beach defenses.

Underwater demolition is a core skill of the SEAL training program. Basic Underwater Demolition/SEAL (BUD/S) training takes up half of the fifty-week-long basic training regime. BUD/S is one of the most physically demanding training courses in the US military. Potential SEALs are pushed to their physical and mental limits. About 60 percent of candidates drop out. Most drop out during a five-and-a-half day period of nonstop training known as Hell Week.

◀◀ A naval EOD technician surfaces after a training exercise on the bottom of the ocean.

18

DEMOLITIONS EXPERTS

Underwater Demolition

SEALs learn to use scuba gear or mini submarines to approach coastlines without being seen. The training involves complex exercises. The candidates learn to swim up to targets without being detected by "defenders." They are trained to place waterproof explosive charges on beach obstacles, harbor infrastructure, or the hulls of enemy warships. The mission is not over once the explosives are in place. SEALs must make a safe escape before demolitions charges are detonated.

EYEWITNESS

"BUD/S prepares us to believe we can accomplish the mission—and to never surrender. No SEAL has ever been held prisoner of war. The only explicit training we receive in BUD/S is to look out for each other—leave no one behind."

—Howard E. Wadsin, former Navy SEAL

A US Navy EOD technician raises his hand as he takes his oath of service underwater.

MILITARY JOBS

▶▶ ASSAULT TEAMS

Using explosives is key to a successful assault on an enclosed space, such as a building. Demolitions experts can blow open doors or windows, or create a hole for an assault team.

An assault on a building or a target such as an airplane or ship is a demanding challenge. An assault force incorporates a skilled demolitions team. The demolitions team leader advises the assault commander on the best place to position charges and how much explosive is needed.

⌄ A demolitions team gets ready to storm a building as an explosive charge blows open a door.

DEMOLITIONS EXPERTS

US Marines use explosives to blow open the door of a house during a training exercise in forced entry.

Two Teams

During an assault, the demolitions team is split into two. The first group goes in with the first wave of troops. It places the demolition charges at the breaching point. The second group detonates the explosives. This group is usually positioned with the troops that will provide covering fire for the assault. That gives them a good view of the unfolding action, so they can detonate the charges at the best time. Timing is often the key to whether an assault succeeds or fails.

IN ACTION

In May 2011, SEAL Team Six landed by helicopter outside the compound in Pakistan where terrorist leader Osama bin Laden was hiding. Demolitions experts breached the compound wall and then blew open the iron doors of the house and created holes in its walls. This part of the operation took only two minutes and the SEALs were then able to enter the house and kill bin Laden.

MILITARY JOBS

▶▶ CLEARING OBSTACLES

Combat engineers are often called in to clear a path for a large-scale assault. This might mean clearing minefields or other obstacles from the battlefield, such as barbed wire or antitank barriers. These tasks require large units of demolitions engineers.

Breaching obstacles may require special explosives. The charges are often put in position while the troops are under fire, so engineers work with armored cars and close-protection troops in a combined-arms operation. Most tactical breaching operations are carried out simultaneously at different locations so large numbers of troops can attack at the same time.

▶▶ Engineers prepare to destroy barbed wire with a Bangalore torpedo, which is a metal tube filled with explosives.

DEMOLITIONS EXPERTS

▶▶ Engineers place a charge, ready to clear a path for an advance.

Into Position

Once breaching points are selected, engineers move into position. The accompanying troops provide covering fire.

If they are crossing a minefield, engineers can fire a cable carrying explosive charges. Otherwise, they place explosive charges next to obstacles. They place markers in the ground to indicate safe pathways for their assault troops.

Once the assault commander receives a signal that all the breaching charges are in place, he warns his troops to take cover and gives the order for the combat engineer to detonate the charges. The assault force can now move forward, with the combat engineers helping guide them through the safe corridor.

EYEWITNESS

"The most effective tool available to the commander is the rehearsal. The complexity of the breaching operation makes rehearsals at every level essential to success. The commander must give his subordinates time to plan how they will execute their assigned mission and time to rehearse the plan with their unit."

—US Army Field Manual

23

MILITARY JOBS

›› IED DESTRUCTION

Terrorist groups often use improvised explosive devices (IEDs). Disarming these weapons is the job of explosive ordnance disposal (EOD) operatives, who never know what they will have to face.

EOD operatives locate suspect devices with special detection equipment that can locate control cables or "smell" explosives. When an IED has been identified, EOD operatives usually use a remote-controlled ground vehicle to approach the device. They make it safe, either by cutting its control wires or firing a shotgun at it to shatter its components without causing a full-scale blast. This not only makes the device safe, but also preserves vital forensic evidence about who made it.

›› **Demolitions experts safely detonate an IED in Iraq.**

DEMOLITIONS EXPERTS

▲ A combat engineer examines an unexploded enemy shell, a common danger in war zones.

Acting Fast

Safe disposal is time-consuming. When there is no time, combat engineers take more rapid action. If there is no danger to human life or critical civilian infrastructure, they place a demolition charge next to the device. This will preferably be done by using a long mechanical arm on an armored vehicle. In extreme situations, a combat engineer simply walks up to the device with the demolition charge. The team takes cover and detonates the charge. The blast either destroys the IED or sets it off. With the device neutralized, the advance can continue safely.

EYEWITNESS

"The enemy will come up with some new way of hiding these IEDs, we'll come up with a new way of defeating it, then they'll move on to a new tactic. We're constantly adjusting."

—Captain John Vogt, Army Engineer, Iraq, 2005

MILITARY JOBS

▶▶ SABOTAGE

Destroying enemy infrastructure such as bridges, railroads, or road junctions requires great expertise. Such structures are usually built to withstand earthquakes and accidents, so engineers need to be able to locate the structure's weak points.

Demolishing a bridge or a building is a specialized engineering task. Demolitions teams need to be methodical. Placing explosives randomly on a structure is unlikely to bring it down. Demolitions teams first have to carry out structural surveys. They drill holes into surrounding concrete or stonework so they can place explosives next to load-bearing steel joists or foundations. Large structures often have

▶▶ Instructed by US engineers, an Afghan Army engineer disarms a simulated IED in 2012.

DEMOLITIONS EXPERTS

multiple stress points that all need to be blown. Explosives are inserted in multiple locations and detonated simultaneously.

Blowing Bridges

Bridges are complex structures. To destroy them, demolitions experts need to collapse the span and destroy any towers or other supports. This prevents enemy engineers simply constructing new spans to cross a gap. It forces them to carry out a time-consuming rebuild operation.

US (left) and Australian engineers check explosive charges set to destroy a bridge during a training exercise.

IN ACTION

The knowledge and expertise of combat engineers is not only useful for destroying bridges. The engineers are also called in if US forces capture an enemy bridge intact. They disarm and dispose of enemy demolitions charges so the bridge can be crossed safely.

MILITARY JOBS

▶▶ UNDERWATER ASSAULT

Demolitions at sea or near the coast are the responsibility of the Navy SEALs. These missions are high risk, however. The SEALs are only sent in if there is no other way to achieve a result.

The advantage of using SEALs is that they can avoid detection. SEAL operatives who swim ashore to a target use scuba gear or mini submarines. They are silent and nearly invisible, and leave no traces behind them.

▶▶ An Explosive Ordnance Disposal technician from the Navy climbs aboard a raft after an exercise.

DEMOLITIONS EXPERTS

One task of SEAL teams is to clear beaches ahead of landings by US Marines. They identify obstacles that could damage or sink landing craft. Without being detected, the SEALs set explosives to clear routes to the beach.

A Navy SEAL approaches the propeller beneath a merchant ship during an exercise in looking for explosive charges planted on vessels.

Destroying Ships

SEAL teams are also trained to swim under enemy warships in harbors. They get past protection nets in the water to get near their targets. They plant explosives on the hulls of the ships. After setting timers, the SEALs swim to safety before the blast.

IN ACTION

In addition to the SEALs, the US Navy also trains dedicated EOD technicians. They are experts at destroying and disarming explosive devices of all types, but can also build and use them. The Navy trains members from the other military services as EOD technicians.

MILITARY JOBS

▶▶ EXPLOSIVES

For a demolitions expert, explosives are among the basic tools of the job. Combat engineers must not only be able to use explosives safely. They must also be able to destroy enemy bombs and arms dumps.

Enemy armies stockpile bombs and artillery shells and insurgents often hide their explosives. If US forces find a large cache of explosives, combat engineers are called in. They make sure the explosives are stable enough to be moved to a safe location. They then bury them before detonating them in a controlled explosion that causes no damage.

▼ **Engineers stack recovered enemy shells before they are buried in sand and safely destroyed.**

30

DEMOLITIONS EXPERTS

A demolitions expert presses a strip of plastic explosive into position as he prepares an IED.

Plastic Explosive

Demolitions experts have a range of explosives and detonators for different missions. The most versatile type is plastic explosive, which is a soft substance like putty that is manufactured in strips or flat panels. It is flexible enough to be used in many situations.

Plastic explosive is small, light, and stable, so it can be carried easily by soldiers. It can be wrapped around bridge supports or fixed to building foundations.

IN ACTION

The US military uses C-4 plastic explosive molded into blocks or strips. Sixteen blocks of C-4 can be put together with four primer assemblies with detonating cords. This creates an M183 "demolition charge assembly." The M183 can be placed in carrying cases for breaching obstacles or demolishing large structures.

31

MILITARY JOBS

HOUSE-ENTRY EQUIPMENT

An assault can lose momentum as forces wait for a building to be breached. Demolitions experts use explosives or brute force to gain entry quickly so the assault can continue.

The key to creating a breach for an assault on a building or other enclosed area is surprise. Demolitions experts might use explosives to blow open doors or windows. Heavily armored doors can be separated from their hinges

US soldiers use a battering ram to force entry into a house in Baghdad, Iraq, during a mission in 2006.

DEMOLITIONS EXPERTS

by carefully placed charges so the door can then be pushed down. If a door is weak, it can be flattened with a battering ram. When there are no usable doors or windows, larger explosive charges are used to blow a breach in a wall.

Quick Action

Once a breach is created, assault troops need to get inside as quickly as possible. That means they need to be close to the explosion. The assault team shelters behind tall blankets or shields only feet away from the blast, then charge through the breach into action.

Marines practice "stacking" behind a blast blanket that protects them within 8 feet (2.4 meters) of a blast.

IN ACTION

One useful form of explosive for breaching doors is detonating cord, or det cord. It is a thin cord filled with high explosives. Although it is intended to be used to detonate other explosives, det cord is powerful enough to be effective on its own in many situations. It can be made into a noose, for example, and used to blast off the handle of a locked door.

MILITARY JOBS

▶▶ OBSTACLE CLEARING EQUIPMENT

If there is time, demolitions experts often use sniffer dogs to locate landmines, which can be disarmed one by one. In the heat of battle it is far quicker to use modern weapons to clear a path through a minefield.

The M58 Mine Clearing Line Charge (MICLIC) is a rocket that shoots a line of explosive charges more than 110 yards (100 meters) into a minefield to clear a path. When the charges detonate, they can blast open a path up to 26 feet (8 meters) wide. The MICLIC rocket launcher is usually mounted on a wheeled or tracked trailer.

▶▶ Soldiers practice firing an M58 MICLIC rocket on a firing range.

34

DEMOLITIONS EXPERTS

In the Minefield

Attached to each meter of the MICLIC cable is 15 pounds (6 kilograms) of C-4 explosive. When the explosive is detonated, any mines nearby blow up with it. The line charges have backup fuses in case the remote detonation system fails. To activate the backup system, the combat engineer has to go into the minefield and manually activate each charge, setting a time delay detonator. Combat engineers dub this dangerous task the "Medal of Honor run."

EYEWITNESS

"[MICLIC] is a tub of C-4 on a high-tension rope with a detonation cord inside. It fires on a rocket over a minefield and is used in counter-landmine warfare to make a lane which trucks can drive through."

—James Gilligan, Combat Engineer, US Marine Corps

A series of explosions mark the detonation of the charges along a MICLIC cable.

MILITARY JOBS

▶▶ IED DISPOSAL EQUIPMENT

Improvised explosive devices (IEDs) are some of the most dangerous weapons in modern warfare. Combat engineers lead the fight against the threat.

The key when dealing with an IED is safety. Combat engineers often do not know how a device has been constructed, how old or decayed it might be, or how stable or unstable it is. Where possible, they first inspect an IED using a robot vehicle fitted with mechanical

▼▼ During training with US forces, an Afghan engineer uses an X-ray device to examine a suspected IED.

DEMOLITIONS EXPERTS

A civilian demolitions expert instructs US troops on how to recognize IEDs in Afghanistan.

arms. Cameras on the robot allow the engineers to observe the device and figure out how to disarm it. A robot vehicle can place a destructive charge against the IED, which is then detonated.

Gathering Information

If an IED seems stable, combat engineers can use X-ray machines to reveal how it was constructed. This information helps the engineers decide how best to destroy the device. It also provides information about how the insurgents made it. This is useful for identifying similar IEDs in the future. It might also give clues to who planted the device.

IN ACTION

Since 2013, US combat engineers have been disarming IEDs with water. The Stingray uses an explosive charge to shoot 40 ounces (1.1 liter) of water in a "blade" that enters the IED at high speed. The water shreds wires, detonators, and any other parts, safely disarming the device without actually detonating it.

MILITARY JOBS

ARMORED VEHICLES

To get to the heart of the battle under enemy fire, combat engineers use a heavily armored tank: the M1 Assault Breacher Vehicle (ABV), or Breacher. Many troops call it simply "the Shredder."

These tracked combat vehicles are variants of the M1 Abrams battle tank. They are modified especially to clear pathways through minefields, as well as clearing roadside bombs and IEDs. The vehicle is protected from

An ABV (right) detonates a mine during mine-clearing operations in Afghanistan in 2011.

DEMOLITIONS EXPERTS

ABVs belonging to the US Marines wait to go into action in Afghanistan in 2011. They led the way in Operation Black Sand, a mission to clear a path through a field of IEDs.

blasts by additional armored tiles on its hull. It can withstand multiple hits and still carry on working. It is armed with a machine gun for protection.

MICLIC

Behind the Breacher's turret is an M58 Mine Clearing Line Charge (MICLIC) rocket launcher. This can be used to blast a safe corridor through minefields. The Breacher is able to carry a number of MICLIC rockets, so it has the capability to clear several minefields one after another.

IN ACTION

The M1 ABV was originally developed by the US Marine Corps. It made its combat debut in Afghanistan. It proved highly successful at blasting routes through insurgent minefields and past IEDs. Army combat engineers were so impressed that they subsequently bought their own fleet of the vehicle for breaching duties.

39

MILITARY JOBS

ZHAWAR KILI CAVES, 2002

After the terrorist attacks on September 11, 2001, US troops led an invasion of Afghanistan. They hoped to eliminate al-Qaeda, the organization that carried out the attacks, and its leader, Osama bin Laden.

Along the border between Pakistan and Afghanistan, al-Qaeda had set up bases and arms caches in the caves that riddled the mountainous region. The caves were one of the primary targets of the invasion. In January 2002, a platoon of US Navy SEALs and other special forces were dropped by helicopter around the Zhawar Kili

>> **US Navy SEALs explore the entrance of one of about seventy caves they found at Zhawar Kili.**

DEMOLITIONS EXPERTS

cave complex. After a brief battle to drive off al-Qaeda fighters, Task Force K-Bar entered the caves. Over the next nine days they recovered documents, weapons, and other material from the network of seventy caves.

SEALs inspect weapons found inside a room in the caves.

EYEWITNESS

"The warriors of Task Force K-Bar established an unprecedented 100 percent mission success rate across a broad spectrum of special operations under extremely difficult and constantly dangerous conditions."
Presidential Unit Citation, 2004

Demolition Charges

The cave complex was remote, with no roads. That made it impossible to remove the hundreds of tons of weapons and ammunition from the caves. It was vital to prevent this arsenal falling back into al-Qaeda's hands when the task force withdrew. The SEALs used their demolitions experts to rig explosive charges throughout the caves. They called in a wave of US aircraft to detonate the explosives with bombs. The caves collapsed, safely sealing in the weapons.

MILITARY JOBS

▶▶ BAGHDAD, 2003

In April 2003, US forces led a coalition invasion of Iraq. Armored columns of the 3rd Infantry Division pushed into the center of Baghdad to seize key positions. They soon came under counterattack.

The troops of the 64th Armored Regiment were soon pinned down under enemy small-arms and mortar fire at a junction named Objective MOE. The Iraqis started driving taxis, cars, and buses packed with explosives at the US positions. The first of these were stopped by American tank fire, but the waves of suicide vehicles kept coming. Demolitions experts then advanced into action.

▶▶ An IED-detecting vehicle passes by as a combat engineer keeps watch in Iraq.

DEMOLITIONS EXPERTS

Smoke rises in southwest Baghdad as demolitions experts detonate an IED spotted by Iraqi civilians along a major route into the city.

Roadblocks

Racing across open ground under heavy fire, the engineers began placing demolition charges under road signs, palm trees, and lamp posts around the junction. The engineers planned to blow up the targets so they fell across the roads. They would create obstacles to stop the Iraqi vehicles getting close to the Americans. US tank crews moved forward to provide cover as the engineers placed their charges and inserted detonators.

The combat engineers pulled back and set off the explosives. The obstacles fell across the roads as planned. Although the battle raged for seventeen hours, the makeshift roadblocks prevented any more suicide bombers getting through.

EYEWITNESS

"This is definitely a job that has hazards. We've taken a lot of factors into account to mitigate that risk. The men are ready to go out there and take care of the hazards."

—Captain John Vogt, Combat Engineer, US Army

MILITARY JOBS

▶▶ HELMAND PROVINCE, 2009

In the late 2000s, Helmand Province, Afghanistan, was one of the most dangerous locations for US military personnel. Taliban insurgents staged ambushes and planted many IEDs.

▽ Marine engineers advance into a simulated minefield to practice for missions in Afghanistan and Iraq.

The Taliban planted hundreds of IEDS in fields around the towns and villages they controlled. They expected the devices to prevent any US attacks.

DEMOLITIONS EXPERTS

A US Army sergeant radios his commanders to report the clearance of anti-personnel obstacles in Afghanistan.

In December 2009, however, the US Marines set out to clear Taliban fighters from the town of Now Zad. The 2nd Combat Engineer Battalion used its Armored Breacher Vehicles (ABVs) for the first time in combat.

Shock Assault

As Marines waited, the ABV fired a Mine Clearing Line Charge (MICLIC) to blast a path through the IED field. Within minutes, columns of Marines were streaming forward into town. The IEDs had failed to hold them up at all. Soon the shocked Taliban fighters were fleeing from the town.

EYEWITNESS

"For me it's an adrenaline rush. Most of the time it was also scary because a lot of us, when we went on a mission, we would ask ourselves, 'We made it back yesterday, are we gong to make it back today?'"

—Kenny Bower, US Army Engineer

45

MILITARY JOBS

GLOSSARY

abseil To descend a vertical surface rapidly using ropes.

amphibious Describes an operation that takes place on sea and on land.

armor Armored vehicles, such as tanks.

arms dump A store of weapons.

breach To create a gap in a wall or barrier.

cache A collection of weapons stored in a hidden place.

coalition A temporary alliance of countries to achieve a goal.

compound A secure area enclosed by a wall.

fuse A cord that is set on fire to trigger an explosive charge.

grappel hook A device with many hooks or claws attached to a rope.

hull The main body of a boat or ship.

infrastructure The basic physical structures of a society, such as roads, bridges, and power lines.

insurgents Rebels fighting against a government or an invasion force.

mine A type of bomb buried just beneath the ground.

minefield An area planted with mines.

ordnance Weapons such as bombs, rockets, or missiles.

psychological Related to the state of someone's mind.

reconnaissance Observation of an enemy's position.

sabotage The destruction of enemy equipment or infrastructure.

simulated Imitating the conditions of battle for training purposes.

scuba Underwater breathing equipment.

stockpile To accumulate a large amount of something, such as weapons.

trip wire A wire stretched close to the ground that sets off an explosion when disturbed.

surveillance Close observation of the enemy.

FURTHER INFORMATION

BOOKS

Carlisle, Rodney P. *Afghanistan War.* America at War. New York: Chelsea House Publications, 2010.

Goldish, Meish. *Bomb-Sniffing Dogs.* Dog Heroes. New York: Bearport Publishing, 2012.

Loveless, Antony. *Bomb and Mine Disposal Officers.* World's Most Dangerous Jobs. St. Catharines, ON: Crabtree Publishing Company, 2009.

Reed, Jennifer. *Marines of the US Marine Corps.* People of the US Armed Forces. Mankato, MN: Capstone Press, 2009.

Tougas, Shelley. *Weapons, Gear, and Uniform of the Iraq War.* Mankato, MN: Capstone Press, 2012.

Wood, Alix. *Explosives Expert.* World's Coolest Jobs. New York: PowerKids Press, 2014.

WEBSITES

www.armystudyguide.com/content/Prep_For_Basic_Training/army_mos_information/combat-engineer-21b.shtml
A private site with details of how to become a combat engineer.

www.goarmy.com/careers-and-jobs/browse-career-and-job-categories/construction-engineering/combat-engineer.html
US Army careers advice for potential combat engineers.

science.howstuffworks.com/military/army-careers/what-does-army-combat-engineer-do1.htm
How Stuff Works pages about being an Army combat engineer.

Publisher's note to educators and parents: Our editors have carefully reviewed these websites to ensure that they are suitable for students. Many websites change frequently, however, and we cannot guarantee that a site's future contents will continue to meet our high standards of quality and educational value. Be advised that students should be closely supervised whenever they access the Internet.

MILITARY JOBS

INDEX

Afghanistan 38, 39, 40–41, 44–45
al-Qaeda 40–41
Armed Services Vocational Aptitude Battery 10
armored vehicles 38–39
Assault Breacher Vehicles 45
assaults 20–21
assaults, underwater 28–29

Baghdad 42–43
Bangalore torpedo 22
Basic Underwater Demolition/SEAL (BUD/S) 18–19
bin Laden, Osama 21, 40
blast blanket 33
breaching 13, 15, 21, 22, 23, 32–33
bridges 27

C-4 explosives 35
combat engineers 4
combined-arms operations 22

D-Day 7
Delta Force 12
demolitions 4, 16–17, 25, 26–27
demolitions, underwater 18–19
detonating cord 33
dogs 15

Engineer Detachment (Canine) 15

engineers, requirements 9
engineers, training 10–11
Explosive Ordnance Disposal (EOD) technicians 18, 19
explosive ordnance disposal 24, 28, 29
explosives 8, 9, 11, 17, 20, 24–25, 30-31
eyewitness 9, 13, 17, 19, 23, 25, 35, 41, 43, 45

Fort Leonard Wood 11

Helmand Province 44–45
history 6–7
house entry 12–13, 32–33

improvised explosive devices (IEDs) 4, 24–25, 26, 31, 36–37, 38, 44
infrastructure 26, 27
Iraq 24, 32, 42–43

killing houses 12
Korean War 7

landmines 14, 15, 34–35

M1 Assault Breacher Vehicle 38, 39
M58 Mine Clearing Line Charge (MICLIC) 34, 35, 39, 45
minefields 9, 14–15, 22, 23, 34–35, 38, 39

Now Zad 45

obstacle clearance 14–15, 22–23, 34–35
Okinawa 6

plastic explosives 31

robot vehicles 36

sabotage 16–17, 26–27
sappers 6
SEAL Team 6 12, 21
SEALs 12, 18, 19, 21, 28–29, 40
selection 10–11
64th Infantry Regiment 42
Special Forces 12
Stingray 37

Taliban 44, 45
Task Force K-Bar 41
terrorism 24–25
training 10–11, 12–13, 14–15, 16–17
training, underwater 18–19
trenches 6, 7

US Army Engineer School 11
US Marine Corps 5, 7

vehicles 38–39

World War I 7
World War II 7

Zhawar Kili Caves 40-41